In summer I wake up early. I ask myself, "Where is the mail?" I play a little game as I wait.

I say, "What will come in the mail?"
Then I answer, "Something good!" I
think of what the mail will be.

Maybe it will be a letter from Fay.

We were in the same classroom.

Maybe it will be a letter from
Uncle Blair. He lives a long way off,
in Greenland. His letters say Airmail
on them.

I play my little waiting game. Soon
Mr. May will bring the mail.

Where does the mail come from?
It can come from anywhere on earth.

Sometimes the wait seems endless!
Will the mail fail to come? Sadly, the
mail may be late.

But look, the mail is here! Mr.
May put it in the mailbox. Thank
goodness! My wait has ended!